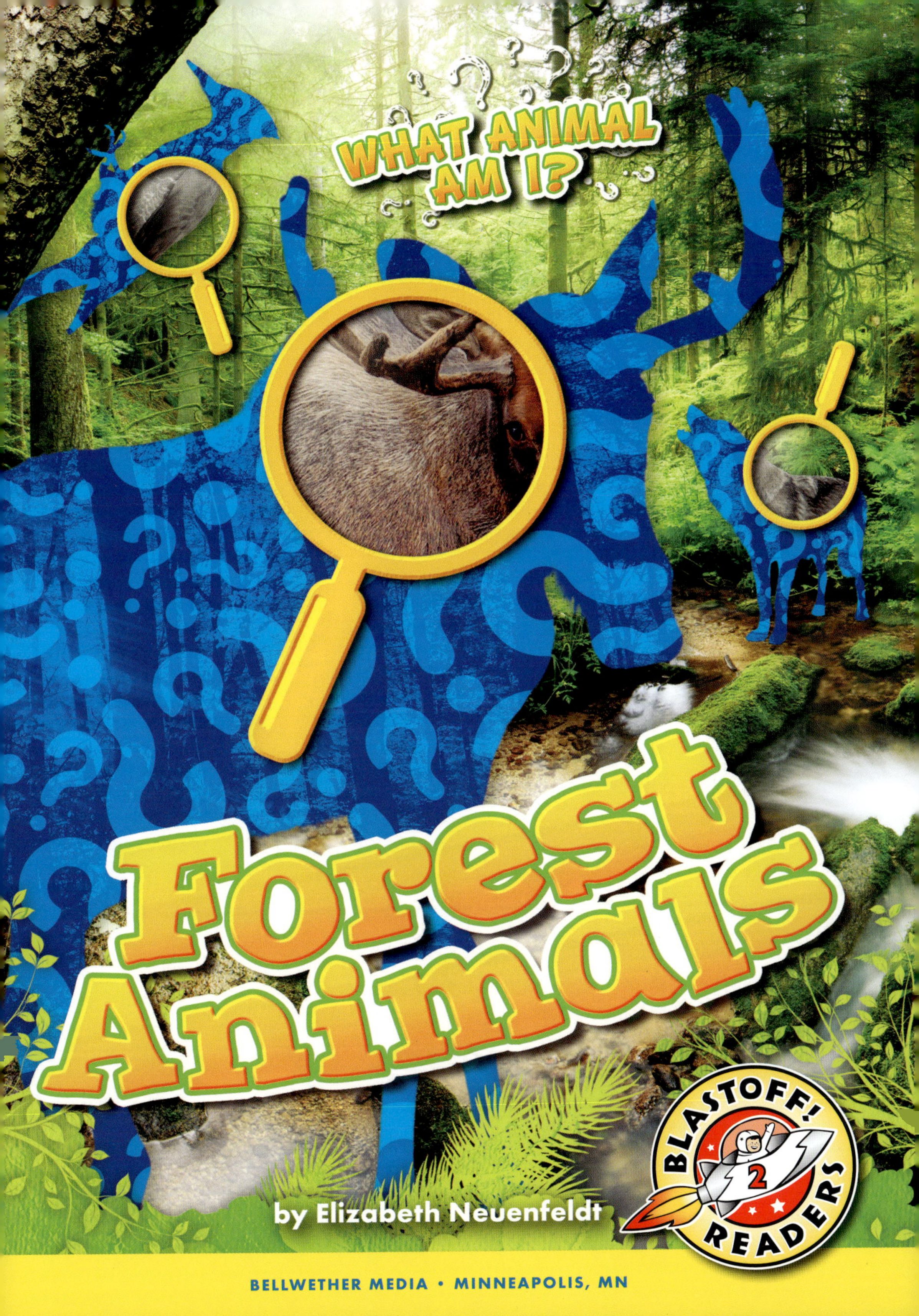
WHAT ANIMAL AM I?
Forest Animals
by Elizabeth Neuenfeldt
BLASTOFF! READERS 2
BELLWETHER MEDIA • MINNEAPOLIS, MN

Blastoff! Readers are carefully developed by literacy experts to build reading stamina and move students toward fluency by combining standards-based content with developmentally appropriate text.

LEVELS

Level 1 provides the most support through repetition of high-frequency words, light text, predictable sentence patterns, and strong visual support.

Level 2 offers early readers a bit more challenge through varied sentences, increased text load, and text-supportive special features.

Level 3 advances early-fluent readers toward fluency through increased text load, less reliance on photos, advancing concepts, longer sentences, and more complex special features.

★ **Blastoff! Universe**

Reading Level

Grade K

Grades 1–3

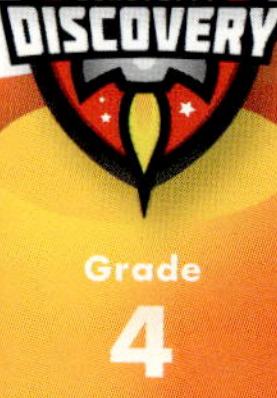

Grade 4

This edition first published in 2023 by Bellwether Media, Inc.

No part of this publication may be reproduced in whole or in part without written permission of the publisher. For information regarding permission, write to Bellwether Media, Inc., Attention: Permissions Department, 6012 Blue Circle Drive, Minnetonka, MN 55343.

Library of Congress Cataloging-in-Publication Data

LC record for Forest Animals available at: https://lccn.loc.gov/2022009384

Text copyright © 2023 by Bellwether Media, Inc. BLASTOFF! READERS and associated logos are trademarks and/or registered trademarks of Bellwether Media, Inc.

Editor: Rachael Barnes Designer: Brittany McIntosh

Printed in the United States of America, North Mankato, MN.

Table of Contents

Welcome to the Forest!

Forests are large areas with many trees. **Boreal forests** have short summers and cold, snowy winters.

Temperate forests have more mild weather. Many animals call forests their home!

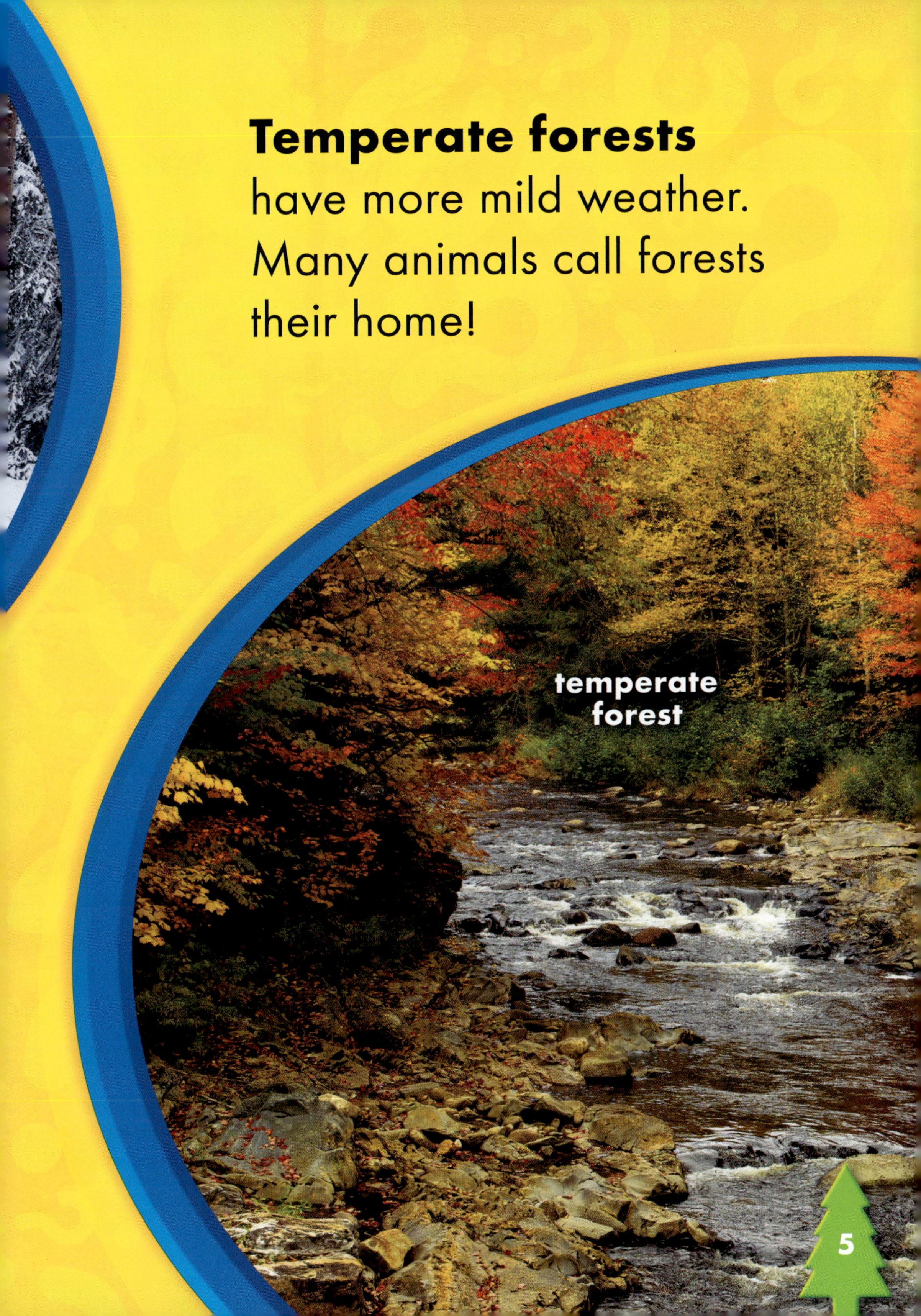

Hopping Around

I am an **amphibian**.
I live on land and in water!

I have wet skin that is usually brown or red. I can croak! What animal am I?

I am a wood frog!
I catch **insects**
and earthworms
with my long tongue.

When it gets cold,
my body freezes.
This helps me live
through winter!

Wood Frog Food
insects
earthworms
spiders

One Big Dog

I am a large **mammal** with thick, gray fur. I have sharp teeth.

I can **howl**. This helps me talk to my **species**. What animal am I?

Gray Wolf Food
deer
sheep
hares
pack

I am a gray wolf!
I live in a **pack**
with other wolves.

We work together
to hunt large animals.

A Feathered Friend

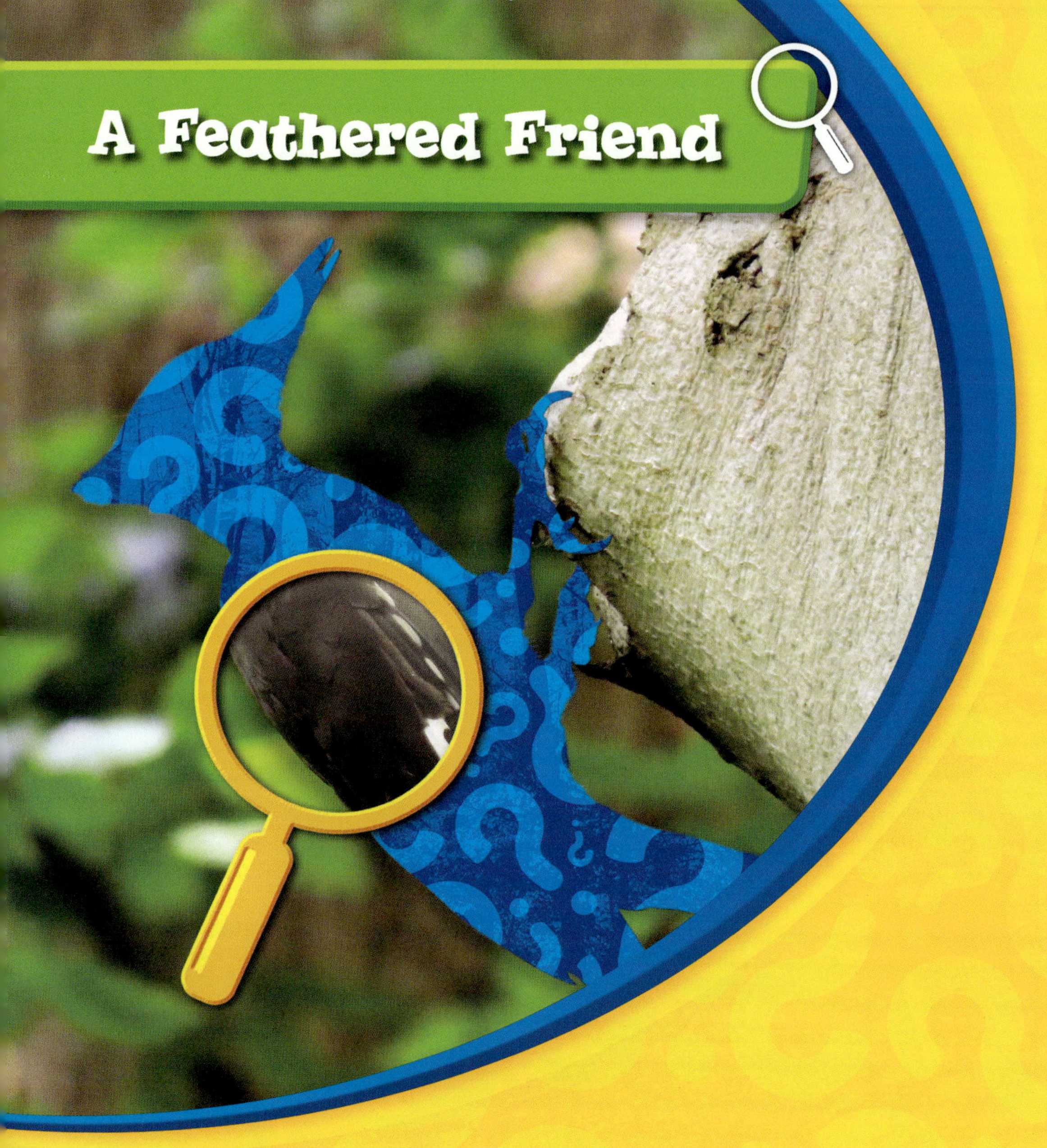

I am a bird with big wings. I have black, white, and red feathers.

I peck trees with my strong beak to build nests. What animal am I?

I am a pileated woodpecker! I eat insects, nuts, and berries.

My long tongue helps me reach insects deep inside trees!

Pileated Woodpecker Food

insects
nuts
berries

A Gentle Giant

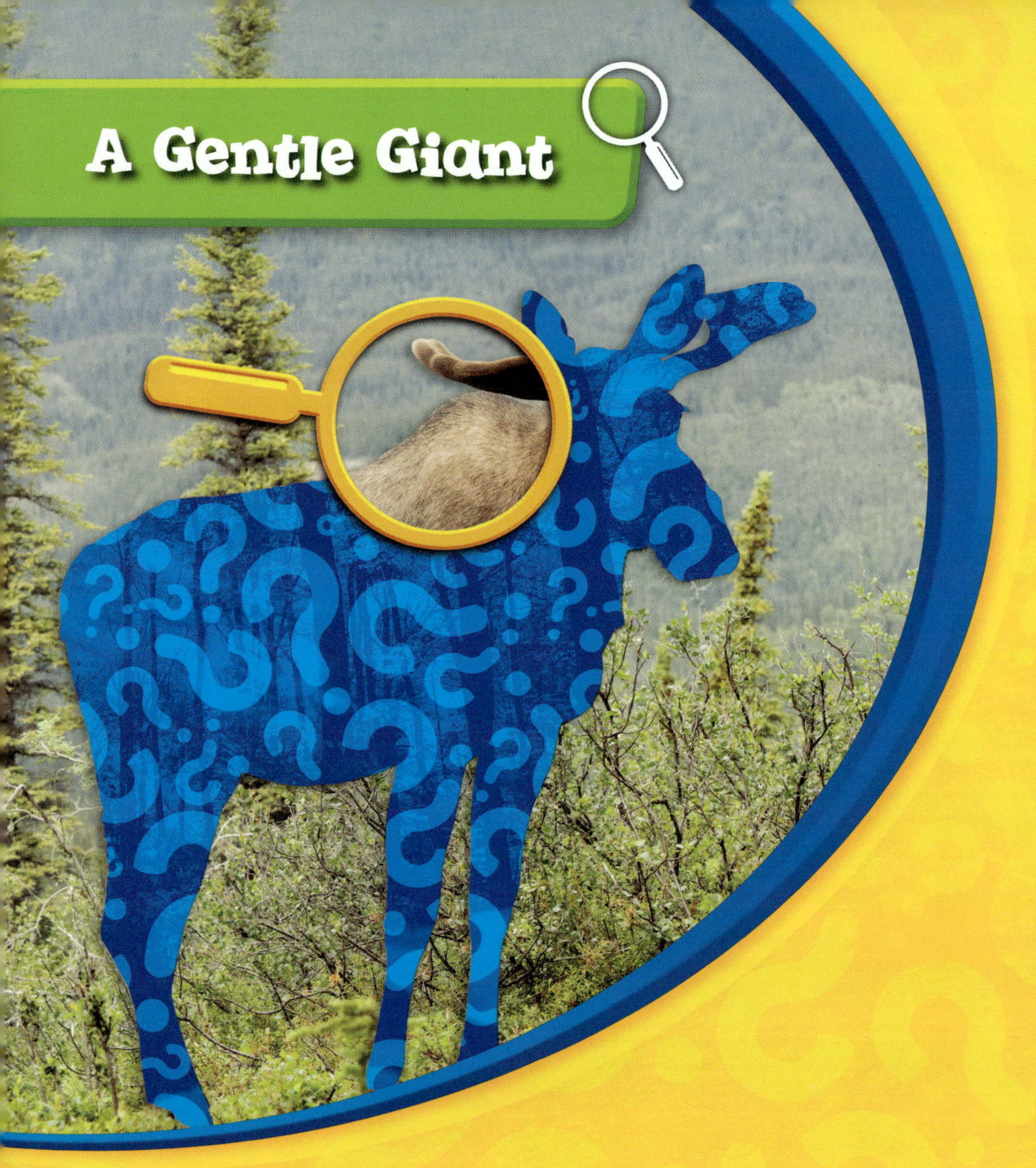

I am a big mammal with dark fur and **split hooves**.

I am a male with huge, flat **antlers**. What animal am I?

Moose Food
shrubs
evergreen trees
pondweed
antlers

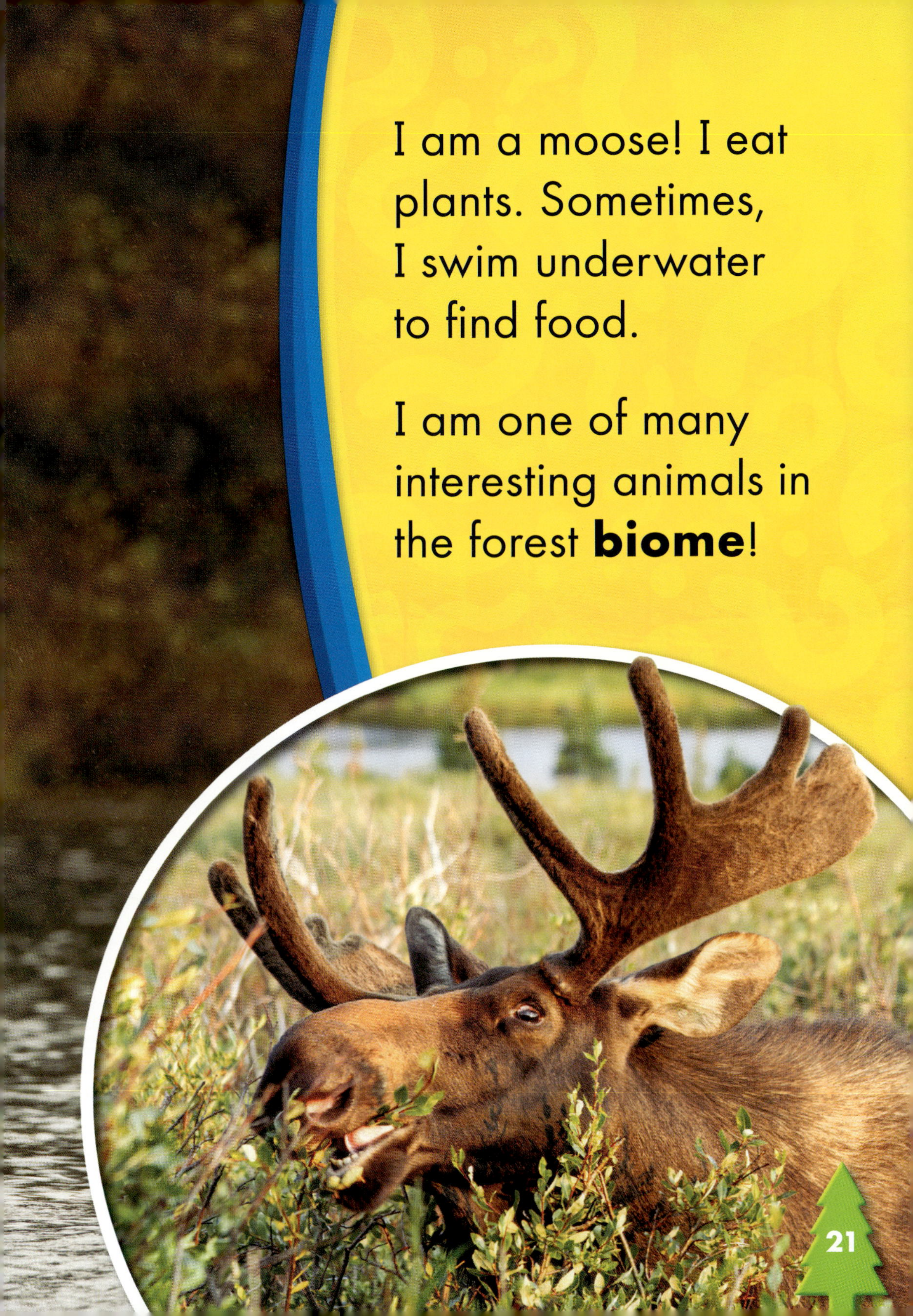

I am a moose! I eat plants. Sometimes, I swim underwater to find food.

I am one of many interesting animals in the forest **biome**!

Glossary

amphibian—an animal that can live on land and in water

antlers—bones that grow on the heads of some animals

biome—a large area with certain plants, animals, and weather

boreal forests—forests that grow in cold regions of Earth just south of the Arctic; most trees in boreal forests are evergreens.

howl—to make a long, loud cry

insects—small animals with six legs and hard outer bodies; an insect's body is divided into three parts.

mammal—a warm-blooded animal that has a backbone and feeds its young milk

pack—a group of animals that lives and hunts together

species—a certain type of animal

split hooves—hooves that are split into two toes; hooves are hard coverings that protect the feet of some animals.

temperate forests—forests that grow in mild weather around the world; most trees in temperate forests lose their leaves in the fall.

To Learn More

AT THE LIBRARY

Berkes, Marianne. *Over in the Forest: A Woodland Animal Counting Book*. Naperville, Ill.: Dawn Publications, 2021.

Kenney, Karen Latchana. *Forests*. Minneapolis, Minn.: Bellwether Media, 2022.

Light, Char. *20 Fun Facts About Forest Habitats*. New York, N.Y.: Gareth Stevens Publishing, 2022.

ON THE WEB

FACTSURFER

Factsurfer.com gives you a safe, fun way to find more information.

1. Go to www.factsurfer.com.
2. Enter "forest animals" into the search box and click 🔍.
3. Select your book cover to see a list of related content.

Index

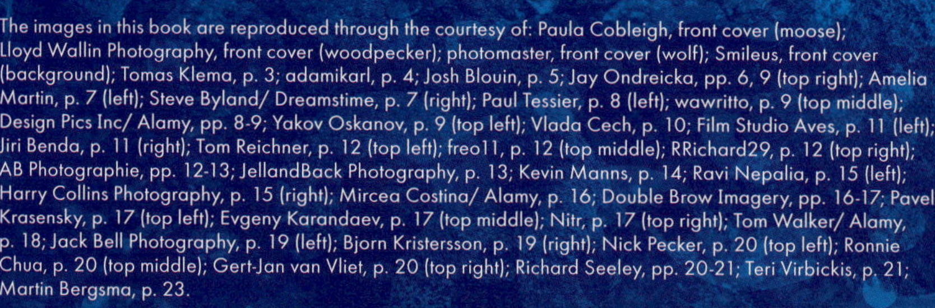

The images in this book are reproduced through the courtesy of: Paula Cobleigh, front cover (moose); Lloyd Wallin Photography, front cover (woodpecker); photomaster, front cover (wolf); Smileus, front cover (background); Tomas Klema, p. 3; adamikarl, p. 4; Josh Blouin, p. 5; Jay Ondreicka, pp. 6, 9 (top right); Amelia Martin, p. 7 (left); Steve Byland/ Dreamstime, p. 7 (right); Paul Tessier, p. 8 (left); wawritto, p. 9 (top middle); Design Pics Inc/ Alamy, pp. 8-9; Yakov Oskanov, p. 9 (top left); Vlada Cech, p. 10; Film Studio Aves, p. 11 (left); Jiri Benda, p. 11 (right); Tom Reichner, p. 12 (top left); freo11, p. 12 (top middle); RRichard29, p. 12 (top right); AB Photographie, pp. 12-13; JellandBack Photography, p. 13; Kevin Manns, p. 14; Ravi Nepalia, p. 15 (left); Harry Collins Photography, p. 15 (right); Mircea Costina/ Alamy, p. 16; Double Brow Imagery, pp. 16-17; Pavel Krasensky, p. 17 (top left); Evgeny Karandaev, p. 17 (top middle); Nitr, p. 17 (top right); Tom Walker/ Alamy, p. 18; Jack Bell Photography, p. 19 (left); Bjorn Kristersson, p. 19 (right); Nick Pecker, p. 20 (top left); Ronnie Chua, p. 20 (top middle); Gert-Jan van Vliet, p. 20 (top right); Richard Seeley, pp. 20-21; Teri Virbickis, p. 21; Martin Bergsma, p. 23.